Bibliographic information published by the German National Library:

The German National Library lists this publication in the National Bibliography; detailed bibliographic data are available on the Internet at http://dnb.dnb.de .

Imprint:

Copyright © 2016 GRIN Verlag, Open Publishing GmbH
Print and binding: Books on Demand GmbH, Norderstedt Germany
ISBN: 978-3-668-13278-8

This book at GRIN:

http://www.grin.com/en/e-book/314634/some-thoughts-about-the-future-of-sailing

Michael Dienst

Some Thoughts about the Future of Sailing

Bionic Science and Yacht Design

GRIN Publishing

GRIN - Your knowledge has value

Since its foundation in 1998, GRIN has specialized in publishing academic texts by students, college teachers and other academics as e-book and printed book. The website www.grin.com is an ideal platform for presenting term papers, final papers, scientific essays, dissertations and specialist books.

Visit us on the internet:

http://www.grin.com/

http://www.facebook.com/grincom

http://www.twitter.com/grin_com

SOME THOUGHTS ABOUT THE FUTURE OF SAILING
Bionic Science and Yacht Design
Berlin, Germany Jan. 2016

Michael Dienst, yachtsman and bionic scientist at the Beuth University of Applied Sciences Berlin, says why biological systems can be a model for future yacht design. The interview on the "Future of sailing" was released in early 2016 in a German sailing magazine[1] in abbreviated form.

Yachting Magazine:
Mr Dienst, you are yachtsman and scientists. Why it might be useful to look to nature for solutions to problems that may occur with sailing?

Mi. Dienst:
Bionics analyses phenomena in nature to derive technical solutions for the future. Especially with the fluidic creatures, animals and plants that live in the water or flying through the air, in the millions of years of biological evolution the nature had deep look into the bag of tricks of physical effects. In fluidics, it depends on smallest amounts of energy, the efficiencies must be high, the losses small and it is an absolute lightweight hip. That's also in sailing that way. And: sailboats must be resilient.

Yachting Magazine: And that means?

Mi. Dienst:
Resilient means to be robust and adaptable over time. Fluidic creatures are under enormous selection pressure. Biosystems that are not optimally adapted to their environment, are dying out. Also marine design can decide over life and death. In a broader sense even beyond the ability of man`s to spread. The entire Pacific Ocean was once populated with sailboats. This only works with resilient constructions.

---

[1] SEGELN MAGAZIN 022016 JAHR TOP SPECIAL VERLAG GmbH
Time for Visions! How we will be sailing in the future? What new technologies are available? Can designers copy ideas from the animal world? We have been asking us in the industry.
Interview: Michael Dienst, Dec. 2015; by Jan Maas.
e-paper: http://www.segeln-magazin.de/februar-ausgabe-2016/4630

Yachting Magazine:
All bodies that move through the water, producing a frictional resistance.
Sharks minimize this resistance with their skin surface. How does this work?

Mi. Dienst:
The panels in the sharkskin are origin teeth! They cover the entire surface of
the shark at different scales, sometimes fine, sometimes coarse, depending on
the location and fluid mechanical requirement. These skin panels are not just
rough, as we think we see and feel, but complex and well structured: in fine
furrows along the flow direction.

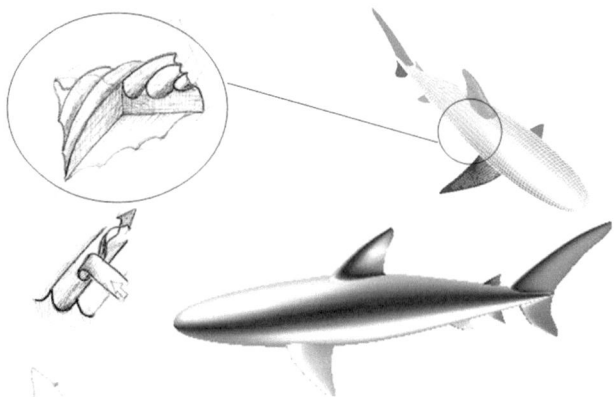

Formation of microvortices into shark skin.

If the near-wall flow velocity shares transversely to the main flow direction, a
micro-vortex rolls up in the furrow, and flows downstream. So the shark skin
turns dangerous cross flow in friendly longitudinal flow. And the thousand
times on the skin. As by a self-generated vortex jacket hurtling the shark at a
rate of up to 80 km/h through the water.

Yachting Magazine:
So you should stick artificial sharkskin on the hull or ribbed structures engrave
the gelcoat?

Berlin, Germany 2016

Mi. Dienst:
Yes absolutely, but it is no longer allowed for racing yachts. With an artificial shark skin, the America's Cup was won by the yacht "Stars & Sripes" in 1987. Then, the artificial shark skin was banned for regatta racing yachts.

Yachting Magazine:
Other animals, such as penguins minimize resistance by their special form. What makes the penguin body so efficiently?

Mi. Dienst:
The secret of the penguin body is the small form-resistance. The penguin body as a "semi-submersible" is unfortunately still poorly explored. But in the late 90s, we have designed a hull modeled on the penguin body for the boatbuilding school in Brake in Germany called: PAVANE. She was beautiful, and as scale fair measurement model also with respect to the characteristic impedance of 13 percent better than all comparative hulls.

Model of the PAVANE for towing test in Brake (Germany). Courtesy J. Borries (1999).

Yachting Magazine: Why this form then not enforced?

Mi. Dienst:
The transmission in a real yacht hull proves to be complicated and fail readily. Cause there are concave surfaces, which boatbuilders do not like. A fellow engineer estimated solely the cost of the construction plans on the scale of a sports car.

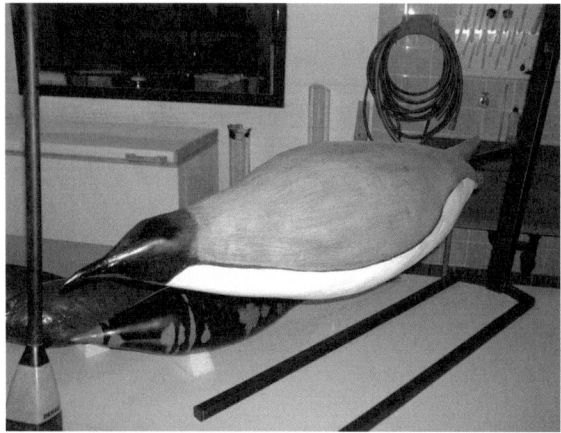

Penguin body (cast and measuring body) FG Bionics and Evolution Technology at the TU Berlin, 1988

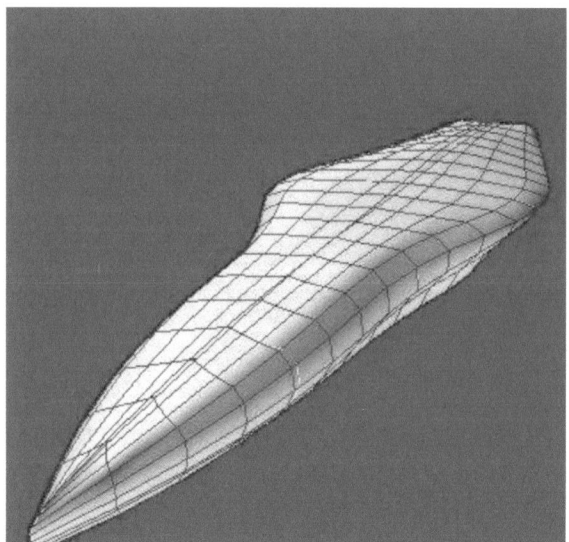

The "PAVANE". Early drafts of a Penguin-Monohull, TU Berlin (about 1996).

Yachting Magazine:
Sailors constantly fighting against fowling on their boats. Dolphins have a skin that cleans itself. How does this work?

Mi. Dienst:
The skin of the dolphins is an elastic membrane comparable. A planar spring-damper-mass-system. Dolphins are able to put their skin in vibrations. Two-dimensional surface waves act flow separation contrary. Put simply, the dolphins "sing" their body surface. This is awesome and efficiently. Perhaps these surfaces also have an influence on the fowling. We do not know exactly. Passive, without singing, silicone surfaces of infestation remain intact against algae in a field trial. The University of Applied Sciences Bremen tested several measuring surfaces in seawater tests. And see, the silicones surfaces had a significantly lower algae and barnacle growth on, as a sound-reflecting surface comparison.

Yachting Magazine:
To reduce the resistance is the one, the other is to increase the advance. What makes bird wings to efficient airfoils and what does that mean for Riggs of sailboats?

Mi. Dienst:
Ever since Manfred Curry had measured Albatros Wings in the wind tunnel of Professor Junkers in Dessau in the 20s, designers should know how the rig of a sailing yacht may look: ... another reason why we have seen Plane rigs last America's Cup.

Yachting Magazine: But why the implementation has taken so long?

Mi. Dienst:
The aesthetics of Curry-Jollen remains unmatched. In the 20s, it was technically ahead of its time. Today it is known that Yacht Design can benefit from the findings of bionics. Troubleshooting on the model of nature exist and have grown beyond the stage of "Freak phase". But the scene is conservative and requires boost.

Yachting Magazine:
Bird wings and fish body can adapt to the conditions by changing its shape. Is it worth the effort?

Mi. Dienst:
Yes. Adaptation is the key slogan today; in yacht design and in fluid mechanics at all. Adaptive flow components, we call it "**i-mech**", so intelligent mechanism modeled like in living nature, is definitely the design of the control plane surfaces of seagoing vessels of the future. Flow adaptive and flexible. If we manage to implement the body deformation into flow loaded structures, we will minimize resistance and energy loss.

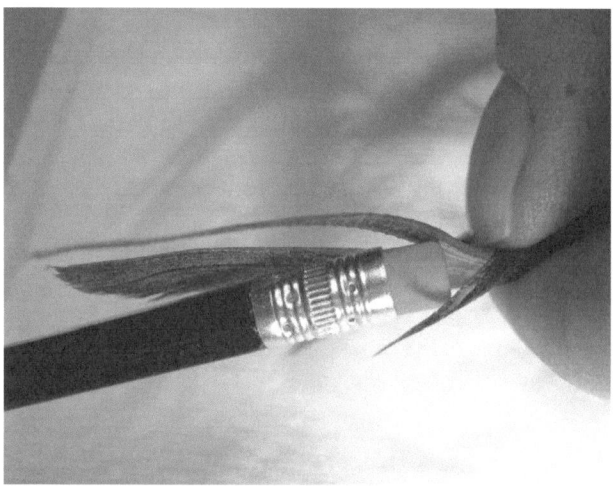

Intelligent mechanics of mackerel fin (Mi. Dienst, 2006)

Yachting Magazine:
Can an efficient sailing style derive from flying and swimming creatures?

Mi. Dienst:
Yes, of cause. The so-called "Gradient Sailing" of the albatross is certainly a good instruction manual for kite surfing.
Sculling? The transfer of the flapped flight of birds is prohibited in all race classes. And I add: Because it is so efficient.
Exploiting vortex structures in air and water? That sounds futuristic, but that it is already researched. It only requires suitable sensors. Here the living nature provides plenty of templates: the whiskers of the sea otter, for example.

Yachting Magazine: Finally. Is there a secret tip from nature?

Mi. Dienst:
Research will take place today not secretly held, but publicly. Open Design; it's good. Two years ago, we have flow adaptive artificial beaver patented as a reactive surface for guidance and control wing. The surface system has guard hairs and works a bit like feathers and sharkskin in one.
But I talk and talk about it ...

Yachting Magazine: .. and?

Mi. Dienst:
The artificial beaver is available in black, white and tabby. But so far there was no one to raise his rudder blade. That's weird.

Yachting Magazine: Many thanks.

Michael Dienst lives and works in Berlin and is sailing for the Club Nautique Francais de Tegel (CNFT). He is spokesman at the BIONIC RESEARCH Unit at the University of Applied Sciences Berlin and lecturer for Bionic Engineering at the Industrial Design Institute of the University of Applied Sciences in Magdeburg.

The 20er Curry-Jolle GER 336 build 1928. Heide, Michael and a goddess from the future. (Courtesy J. Kiewert, at VSAW Berlin, 2014)